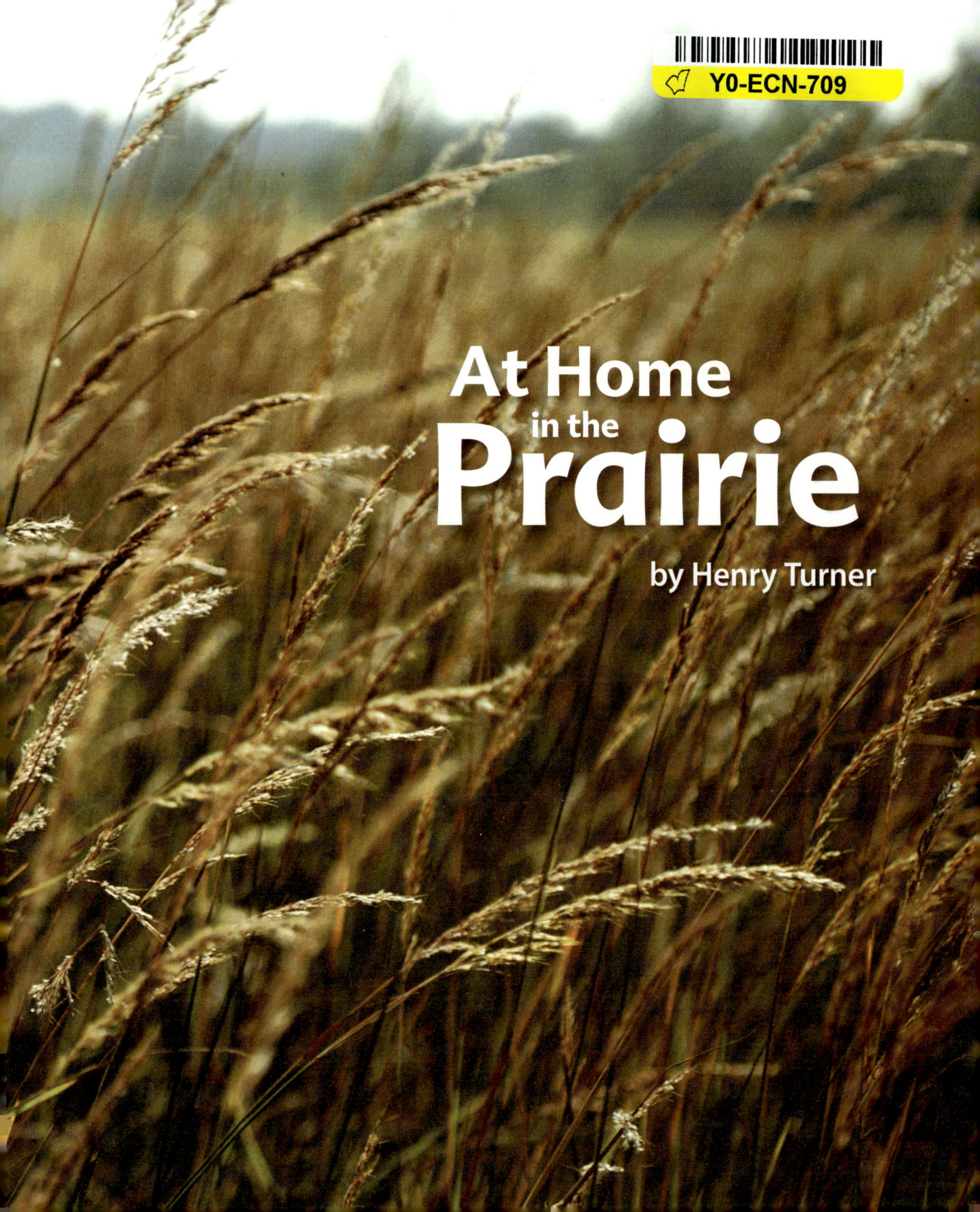

# At Home in the Prairie

by Henry Turner

# Contents

Science Vocabulary. . . . . 4

What Is a Prairie?. . . . . . . 8

What Plants and
Animals Need. . . . . . . . . 14

Prairie Habitats. . . . . . . . 24

Conclusion . . . . . . . . . . 28

Share and Compare . . . . . 29

Science Career . . . . . . . . 30

Index . . . . . . . . . . . . . 32

# Science Vocabulary

**Earth**
Earth is the planet on which we live.

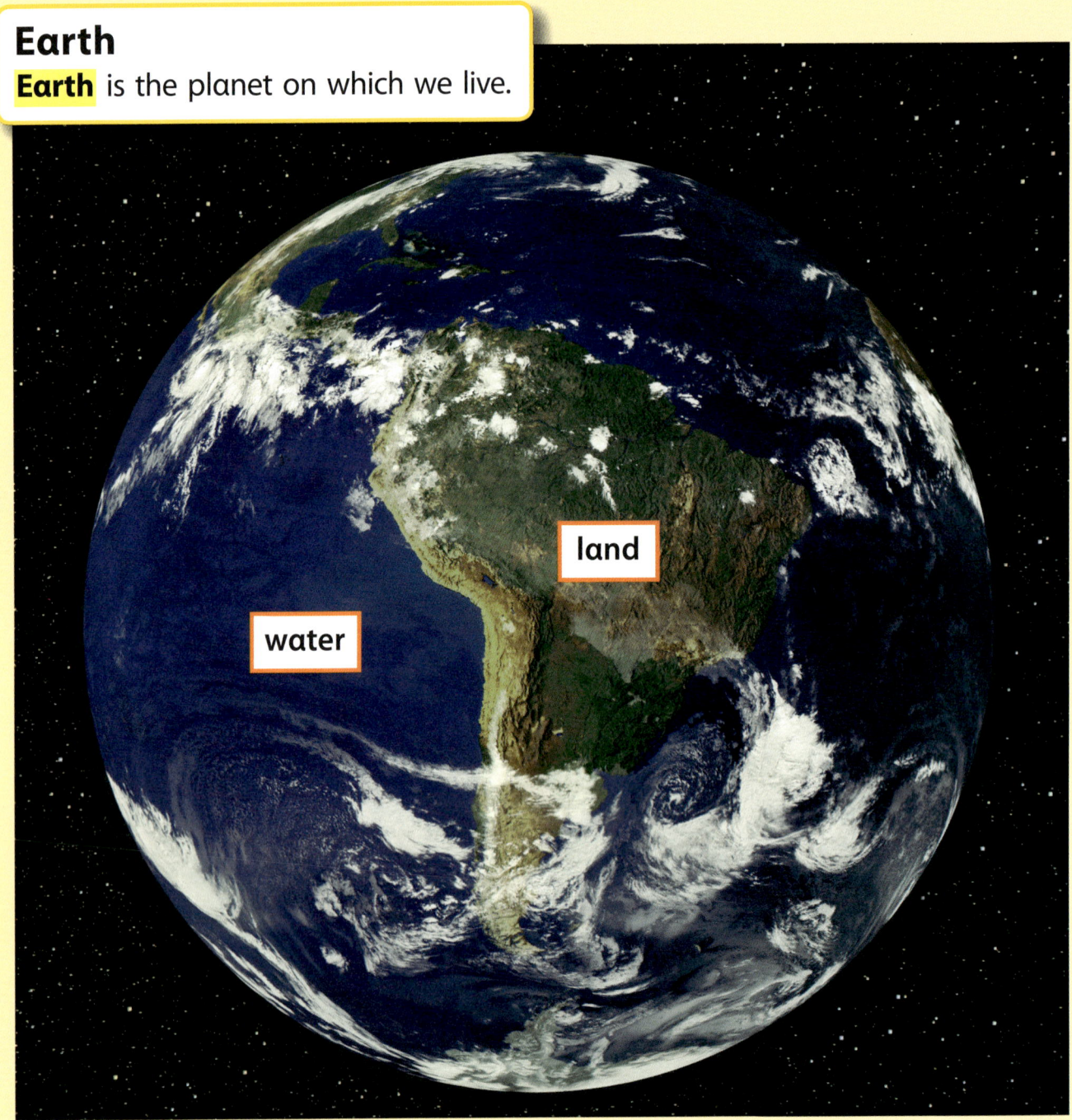

Earth has large areas of land and water.

## habitat
A **habitat** is a place where living things can get what they need to stay alive.

prairie

The prairie is a **habitat.**

## survive
When living things **survive**, they get what they need to stay alive.

This prairie hawk hunts for food. It needs food to **survive.**

**oxygen**
**Oxygen** is a gas in air and water.

This ferret breathes in **oxygen.**

**shelter**
A **shelter** is a safe place where a living thing can make its home and grow.

This nest is a **shelter** for the eggs.

## nutrients
**Nutrients** are parts of food and soil. They help living things stay healthy and grow.

**My Science Vocabulary**

| Earth |
| energy |
| habitat |
| nutrients |
| oxygen |
| shelter |
| survive |

grass

soil

**Nutrients** in soil help grass grow.

## energy
**Energy** is the ability to do active things.

Sun        Plant        Mouse

A plant gets **energy** from the sun. The mouse gets energy when it eats the plant.

7

# What Is a Prairie?

**Earth** has many **habitats**. A prairie is a land habitat.

**Earth**

**Earth** is the planet on which we live.

**habitat**

A **habitat** is a place where living things can get what they need to stay alive.

A prairie has few trees. It has a lot of grass!

Some plants and animals live in a prairie.

Indian paintbrush plants

The prairie is their habitat. The prairie is their home.

Prairie dog

These plants and animals have special parts.

Some animals stay away from this bull thistle's sharp thorns.

These special parts allow them to **survive** in the prairie.

This prairie hawk uses its sharp claws to hunt.

**survive**

When living things **survive,** they get what they need to stay alive.

# What Plants and Animals Need

Animals must have air, water, food, and **shelter** or space. Plants must have air, water, light, and space.

Plant roots take in water from soil.

**shelter**

A **shelter** is a safe place where a living thing can make its home and grow.

Animals must have plants or other animals for food.

Bison find food and water in the prairie.

All animals must have **oxygen.** Plants give off oxygen.

Fox

Ferret

Elk

**oxygen**

**Oxygen** is a gas in air and water.

Some animals use plants for shelter. Some animals use plants to make nests.

Animals help plants, too. They loosen soil. Plants can grow better in loose soil.

This worm loosens soil as it moves.

Animals carry seeds to new places. New plants can grow in those places.

seed

This bison is carrying seeds. How do you think they got on its fur?

All plants and animals die. But they are still part of the prairie.

This is the skull of an animal that died.

Their bodies decay, or rot. This adds **nutrients** to the soil.

New plants grow in a prairie.

**nutrients**

**Nutrients** are parts of food and soil. They help living things stay healthy and grow.

21

A prairie has many plants and animals. They need each other. Plants get **energy** from the sun.

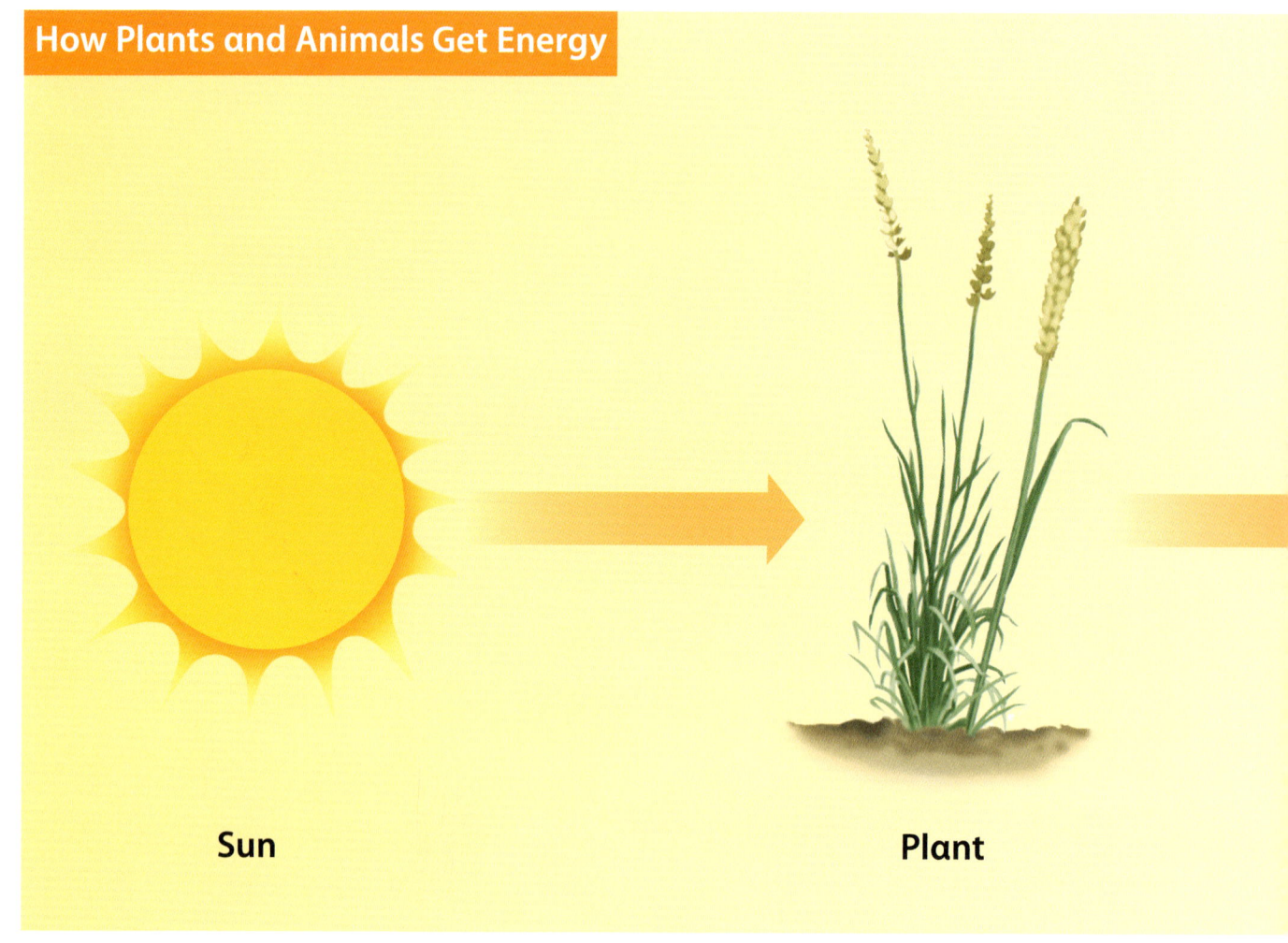

**energy**

**Energy** is the ability to do active things.

Animals get energy from the food they eat. The plant is food for the mouse. The mouse is food for the hawk.

# Prairie Habitats

Plants and animals live in a prairie. Some can't live in other habitats.

Yet people plow prairie land to plant crops. So there is less open prairie land.

Some plants and animals need the open space of prairies. They must have room to live and grow.

# Conclusion

Many plants and animals live in prairies. These plants and animals get what they need to survive in their prairie habitat. Some of these plants and animals can't live in other habitats.

## Think About the Big Ideas

1. Where do plants and animals live?
2. What do prairie plants and animals need to survive?
3. How do prairie plants and animals depend on each other?

# Share and Compare

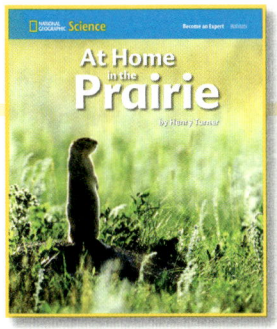 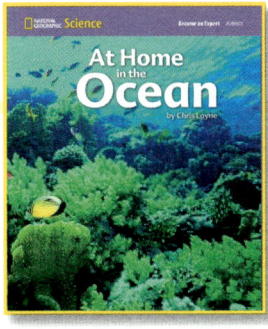 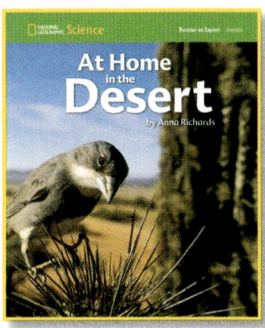

### Turn and Talk

Compare the habitats in your books. How are they different? How are they alike?

### Read

Find your favorite part of the book and read it to a classmate.

### Write

Tell what plants and animals need to survive. Share what you wrote with a classmate.

### Draw

Show an animal or plant in its habitat. Share your drawing with a classmate.

## Science Career

### Meet Greg Marshall

Scientists ask a lot of questions. Sometimes technology can help them answer these questions.

Greg Marshall wanted to see animals in their habitats. So he invented a camera and put it on different kinds of animals.

It took him many tries to get the camera exactly right. He called his invention *Crittercam*.

Crittercam throughout the years

1987

1991

# Index

Earth . . . . . . . . . . . . . . . . . 4, 7–8

energy . . . . . . . . . . . . . . 7, 22–23

grass . . . . . . . . . . . . . . . . . . . 7, 9

habitat . . . . . . 5, 7–8, 11, 24, 28–30

nest . . . . . . . . . . . . . . . . . . . 6, 17

nutrients . . . . . . . . . . . . . . . 7, 21

oxygen . . . . . . . . . . . . . . . . 6–7, 16

prairie . . 5, 8–11, 13, 20, 22, 24–26, 28

shelter . . . . . . . . . . . . . 6–7, 14, 17

survive . . . . . . . . . . . 5, 7, 13, 28–29

water . . . . . . . . . . . . . . . . . . 4, 14

32

**Acknowledgments**
Grateful acknowledgment is given to the authors, artists, photographers, museums, publishers, and agents for permission to reprint copyrighted material. Every effort has been made to secure the appropriate permission. If any omissions have been made or if corrections are required, please contact the Publisher.

**Photographic Credits**
Cover (bg) Todd Korol/Aurora Photos; Cvr Flap (t), 5 (t), 8–9 Niv Koren/Shutterstock; Cvr Flap (c), 15, 28 José Enrique Molina/age fotostock; Cvr Flap (b), 21 Shanin Glenn; Title (bg) James P. Blair/National Geographic Image Collection; 2–3 Joel Sartore/National Geographic Image Collection; 4 NASA/Corbis; 5 (b), 13 Michael Melford/National Geographic Image Collection; 6 (t), 16 (tr) Dmitry Grivenko/Shutterstock; 6 (b), 17 Joel Sartore/National Geographic Image Collection; 7 (t) 14 Joel Sartore/Grant Heilman Photography; 10 James P. Blair/National Geographic Image Collection; 11 Bates Littlehales/National Geographic Image Collection; 12 Peter Barrett/Shutterstock; 16 (tl) Frank Mathers/Shutterstock, (b) Alan Scheer/Shutterstock; 18 chudoba/Shutterstock; 19 Mark Newman/Bruce Coleman Inc./Photoshot; 20 Danita Delimont/Alamy Images; 24–25 Dean Conger/National Geographic Image Collection; 26–27 M. Williams Woodbridge/National Geographic Image Collection; 30–31 (bg) Annie Griffiths Belt/National Geographic Image Collection; 30–31 (insets) Greg Marshall/ National Geographic Remote Imaging; 31 Peter McBride; Inside Back Cover (bg) LOOK Die Bildagentur der Fotografen GmbH/Alamy Images.

**Illustrator Credits**
7 (b), 22–23 John Kurtz

Neither the Publisher nor the authors shall be liable for any damage that may be caused or sustained or result from conducting any of the activities in this publication without specifically following instructions, undertaking the activities without proper supervision, or failing to comply with the cautions contained herein.

**Program Authors**
Randy Bell, Ph.D., Associate Professor of Science Education, University of Virginia, Charlottesville, Virginia; Malcolm B. Butler, Ph.D., Associate Professor of Science Education, University of South Florida, St. Petersburg, Florida; Kathy Cabe Trundle, Ph.D., Associate Professor of Early Childhood Science Education, The Ohio State University, Columbus, Ohio; Nell K. Duke, Ed.D., Co-Director of the Literacy Achievement Research Center and Professor of Teacher Education and Educational Psychology, Michigan State University, East Lansing, Michigan; Judith Sweeney Lederman, Ph.D., Director of Teacher Education and Associate Professor of Science Education, Department of Mathematics and Science Education, Illinois Institute of Technology, Chicago, Illinois; David W. Moore, Ph.D., Professor of Education, College of Teacher Education and Leadership, Arizona State University, Tempe, Arizona

**The National Geographic Society**
John M. Fahey, Jr., President & Chief Executive Officer
Gilbert M. Grosvenor, Chairman of the Board

Copyright © 2011 The Hampton-Brown Company, Inc., a wholly owned subsidiary of the National Geographic Society, publishing under the imprints National Geographic School Publishing and Hampton-Brown.

All rights reserved. No part of this book may be reproduced or transmitted in any form or by any means, electronic or mechanical, including photocopying, recording, or by an information storage and retrieval system, without permission in writing from the Publisher.

National Geographic and the Yellow Border are registered trademarks of the National Geographic Society.

National Geographic School Publishing
Hampton-Brown
www.NGSP.com

Printed in the USA.
RR Donnelley, Johnson City, TN

ISBN: 978-0-7362-7581-1

11 12 13 14 15 16 17

10 9 8 7 6 5 4